The Red Poll Cattle and Farm Conditions
Published in the Interest of Red Poll Cattle Breeders

by Ohio Red Poll Breeders' Association

with an introduction by Jackson Chambers

This work contains material that was originally published in 1917.

This publication is within the Public Domain.

This edition is reprinted for educational purposes
and in accordance with all applicable Federal Laws.

Introduction Copyright 2018 by Jackson Chambers

Self Reliance Books

Get more historic titles on animal and stock breeding, gardening and old fashioned skills by visiting us at:

http://selfreliancebooks.blogspot.com/

Introduction

I am pleased to present another title in the "Cattle" series.

The work is in the Public Domain and is re-printed here in accordance with Federal Laws.

As with all reprinted books of this age that are intended to perfectly reproduce the original edition, considerable pains and effort had to be undertaken to correct fading and sometimes outright damage to existing proofs of this title. At times, this task is quite monumental, requiring an almost total "rebuilding" of some pages from digital proofs of multiple copies. Despite this, imperfections still sometimes exist in the final proof and may detract from the visual appearance of the text.

I hope you enjoy reading this book as much as I enjoyed making it available to readers again.

Jackson Chambers

List of Members

(January, 1917)

H. C. Price, Newark, O.
J. J. & R. C. Krantz, Dover, O.
F. Nelson, London, O.
Henry S. Kelley, Geneva, O.
Jacob Overly, Bainbridge, O.
Wm. K. Hirshberger, Lancaster, O.
H. E. & J. E. Wynkoop, Eldorado, O.
Frank Hartline, Strasburg, O.
J. W. Lee, Racine, O.
Stump & Etzler, Convoy, O.
H. W. Rinkert, West Liberty, O.
E. M. Kroft, Mt. Perry, O.
M. A. Page & Son, Dennison, O.
G. L. Roush, Springfield, O.
R. O. Evans, Blanchester, O.
Lewis Rodgers, Good Hope, O.
N. R. Peffley, Germantown, O.
Elmer E. John, Dayton, O.
Geo. H. Smith, C. C. Cushman, Mgr., Chillicothe, O.
R. C. Wise, Newark, O.
E. G. Norton, Seville, O.
R. G. Bradfield & Son, West Jefferson, O.
Frank H. Hawley, Le Roy, O.
A. S. Bolen, Fremont, O.
R. H. Statler, Shelby, O.
F. M. Borst, Bainbridge, O.

◻

Directors

Homer C. Price
Frank Hartline
Frank Nelson
J. Wilbur Lee
Stump & Etzler
J. J. and R. C. Krantz

Dual Cows of other Breeders there are--as freaks. But dual type by nature and in execution as a Breed there is but One--that one Breed--The Red Poll.

The Red Poll and Farm Conditions

It is the Law of Average, whatever the field of human endeavor, that sustains.

This very law, applied to Farm Conditions of today, **imperatively** demands that the Cow producing the Milk on the farm must also produce the Feeders and the Beef.

It is this same Law of Average that, while imperatively demanding a dual- nature and type in the Farmers' Cow, has also at the same time, through **hundreds of years** of application and test, proved the Red Poll to be the most truly dual-purpose animal.

Let us now look into this dual, or two sided, Milk and Beef nature of the Red Poll. If the Red Poll is of true dual type and nature it must, at all times, not only produce milk with the best of the milkers but it must, at the same time, produce Beef with the best of the Beef Breeds. Can the Red Poll do this? It can, second to none.

In Milk Production the Red Poll stands fourth in the list in competition with the world. Surpassing even some of the known strictly Dairy Breeds. Beauty No. 31725, one of the

The Red Poll and Farm Conditions

Record Cows of the Breed, has an official record of 20280 pounds of milk and 891.5 pounds of butterfat. Her weight is 1750 pounds and but for her under line is an ideal, typically formed Beef type animal.

Pear No. 24888, weight 1440, is the World's Champion Long Distance Red Polled Cow and is the champion Long Distance Record Cow, over all breeds, in the great Dairy State of Minnesota. Minnesota has one herd of thirty Red Poll Cows with not a Cow in it whose official record is under 460 pounds of butter fat. Absorb this fact and can you then ask, Is the Red Poll a Milker?

In Milk Production the Red Poll surpasses the Jersey, Brown Swiss, Shorthorn and all Guernsey records excepting that of Murue Cowan.

Now let us look into the Beef production of the Red Poll and see how really and truly is the Red Poll nature a dual nature. Here let me say regarding the Dual nature. Such is not a strange or unnatural attribute. It is but the embodiment and perpetuation of God's perfect laws in full response to nature's demands of a balance. Mammary development

The Red Poll and Farm Conditions

at the expense of constitution and vigor, or the ability to lay on fat or weight at the expense of mammary worth, is not God's work. He did not do this. Man did it. Man often thinks he can lift himself by his own boot straps, but he never succeeds in doing so. For this very reason, that the Red Poll is a natural, physically balanced animal and has so been bred, coeval with England's earliest history down to the present day, is a guarantee to the Farmer that it is a Feeder and Beef animal. There is none superior for the Farmer and cornbelt. This statement is fully substantiated by the records of the past ten or more years of the Slaughter Contest at the International of Chicago. Here, as individuals, the Red Poll Feeders have had to show and compete with the Agricultural Colleges. But no year has passed that has witnessed the Red Poll outside the money.

At Smithfield Club Show of England in 1890 a Red Poll Steer dressed 73.72%. This, according to the London Live Stock Journal, has only once been exceeded in England and **never by a full blood steer of any breed.**

Thus in this same law of averages,

The Red Poll and Farm Conditions

cited above, are more than ample grounds to prove the Red Poll as a **Milker** and a **Beef Animal** par excellence. A true, dual type by nature and in execution. Then why be content with any Breed that can not do as well? Or with a breed that can only do one-half as much? Be that half Milk, or be that half Beef? A better cross than the Red Poll I think does not exist.

Seldom is it that the Farmer can, or is called upon to specialize. He must carry forward, at one and the same time, the many diversified features of the farm. Not solely must he look to Sheep, to the Hogs, Brood mares, Milk, Butter, or the Cattle as feeders with which to fill his feed lots. He must keep his eye on and care intelligently for all. As a single unit, must he carry forward his whole scheme of farm efforts. Not all his eggs in one basket, but many eggs in all the baskets is the true farmer's aim and effort. The immutable law of averages compels him to feed, if he would be fed. His land, above all, must be fed if his Family, his Crops and his Bank Account are to be fed. Long, long ago has he learned that

The Red Poll and Farm Conditions

Manure is King when he is in pursuit of land enrichment. Equally as long ago has he learned that a large margin of profit is contained in the roughage and by products of his crops. At that time he also learned that Cattle can turn this roughage and waste of the farm into profit better, quicker and cheaper than anything else. Therefore, as the average Farmer is by nature restricted to certain definite lines and must, perforce, diversify efforts as well as crops, let us examine the qualifications of the Red Poll when subjected to average farm conditions. Easy would it be to appeal to the facts of the Breeds history, long patent to the Red Poll student. But it is my intention to take you outside the authentic channels of the Red Poll history.

The Red Poll Breeder and Student is as desirous of knowing he is **keeping** the middle of the bovine road, as you are in seeking and finding it. Hence any facts bearing on the subject and coming from extraneous sources, from other than dyed in the wool Breeders, are most welcome facts indeed. Thus it is with pleasure I append, verbatim, an article from the

The Red Poll and Farm Conditions

pen of Mr. W. J. Kennedy, of the Iowa Agricultural College. This appeared in the Breeders' Gazette of Feb. 2, 1914, and was entitled BUTTERFAT PRODUCTION UNDER FARM CONDITIONS.

To The Gazette.—We read a great deal about the phenomenal butterfat records made by cows of the respective dairy breeds. So much publicity is given these feats that the beginner is oftentimes led to believe that about all that is necessary to insure success in the dairy business is the purchase of a few cows of this or that breed. It is easy to figure that a few cows producing from 600 to 900 pounds of butterfat each would support a fair sized family in comparative luxury. So much for the theory, but what are the actual facts?

These wonderful feats are interesting and useful in that they show the possibilities of doing unusual things, when the surroundings are all favorable and high records, not economical returns, are the chief consideration. With our high-priced land, feed and labor the cornbelt farmer will be compelled to pay more attention to the dairy end of the cattle business in the

The Red Poll and Farm Conditions

future. Some will make a specialty of dairying. For such men a special purpose dairy breed should be used. It has long since been demonstrated that the good special purpose dairy cow, when given the proper feed, care and management, is one of the most economical machines known to man for converting roughage and concentrates into food products. Other men, and they compose a large constituency, do not care to make a specialty of dairy farming. They wish to do some milking in connection with beef production. Some of these men will milk their cows and rear the calves on skimmilk and grain. Others will milk about one-half of their cows and allow the other half to suckle two calves each. This practice prevails on the highest-priced lands of England and Scotland and has proved to be very profitable.

I have been making a rather careful study of the dairy test association work which is being conducted by the extension department of Iowa State College. The work has grown from year to year and has been very helpful to farmers of Iowa. In 1909 two associations were organized in Black-

The Red Poll and Farm Conditions

hawk county. The two assoications contained 46 herds with a total of some 688 cows. The work commenced on June 11, 1909. A man was placed in charge of each association. He did the testing and tabulated the daily feed an dmilk records kept by the farmers. The results of the first year's work show just what is happening under average farm conditions. These men did not know whether their cows were profitable or not. They wanted to know the truth about each cow and were much interested in the final outcome.

Out of the 688 cows in the two associations 505 completed the full twelve months. They were of the following breeding: 7 Shorthorns, 218 grade Shorthorns, 19 Holsteins, 56 grade Holsteins, 6 Guernseys, 25 grade Guernseys, 3 Jerseys, 37 grade Jerseys, 16 grade Herefords, 3 grade Red Polls, 1 grade Angus, and 114 of mixed breed. Out of these 505 cows 32 made 300 pounds or better of butterfat. In breeding they were as follows: 1 Shorthorn, 16 grade Shorthorns, 5 grade Jerseys, 4 grade Holsteins, 1 Guernsey, 3 grade Guernseys, 1 grade Hereford, and 2 grade Red Polls.

That there were some really useful cows in these associations is evidenced by the fact that the ten best cows made the following:

Rank	Breeding	Age—years	Pounds butter-fat.	Cost per pound of butter-fat.	Cost of feed.	Profit
1	Shorthorn-Angus	6	412.8	7.8c.	$32.36	$106.79
2	Grade Shorthorn	9	410.8	11.0c.	45.20	97.39
3	Grade Shorthorn	8	382.5	8.7c.	33.31	96.60
4	Grade Shorthorn	8	373.5	8.9c.	33.31	92.74
5	Grade Shorthorn	10	366.2	8.8c.	32.06	89.90
6	Grade Shorthorn	6	363.2	9.4c.	34.12	90.68
7	Grade Holstein	6	342.6	11.3c.	3.78	77.21
8	Grade Guernsey	7	337.9	9.6c.	32.36	79.40
9	Grade Shorthorn	8	333.8	9.7c.	32.36	77.29
10	Grade Jersey	3	326.5	9.8c.	31.86	76.56

In figuring the cost of a pound of butterfat only the feed is charged; no allowance is made for labor or interest on investment. The cows were credited with the number of pounds of butterfat produced each month at the prevailing market price. The following table shows that 107 of the 505 cows made 250 pounds of butterfat or better during the 12 months, and gives the details:

No. of cows.	Breeding.	Average lbs. butter-fat per cow.	Average cost per pound butter-fat per cow.	Average cost of feed per cow.	Average profit per cow
*51	Shorthorn	293.0	11.58c.	$33.48	$65.24
*18	Jersey	281.56	14.07c.	39.06	56.05
15	Holstein	284.0	13.18c.	37.09	58.63
12	Guernsey	284.2	17.08c.	47.67	47.34
7	Mixed breeding	259.5	13.20c.	34.30	53.41
2	Grade Red Polls	305.0	8.90c.	26.95	70.76
1	Grade Hereford	311.8	11.00c.	34.20	70.89
1	Grade Angus	267.4	14.20c.	37.96	56.07

*Includes both purebreds and grades. Iowa Agricultural College. W. J. KENNEDY.

The Red Poll and Farm Conditions

Here we find, out of a total of over 600 cows, taking the herds just as they would occur in any county, but three Red Polls and these three were grades. We also find, that of all breeds entered more Red Polls made good, or 300 pounds of butterfat or better, than any other Breed.

We find that—

Out of 225 Shorthorns entered but 17 or 7 1-2% made good.

Out of 75 Holsteins entered but 4 or 5 1-3% made good.

Out of 31 Guernseys entered but 4 or 12.9% made good.

Out on 40 Jerseys entered but 5 or 12 1-2% made good.

Out of 16 Herefords entered but 1 or 6 1-4% made good.

Out of **3 Red Polls entered but 2 or 66 2-3 made good.**

Out of 1 Angus entered none or 00% made good.

Out of 144 Mixed Breed entered none or 00% made good.

But of equal if not greater importance, are the figures of cost of production per pound of butterfat and total profit per animal. Analyzing these figures we find the Red Polls produced a pound of butterfat for **less**

15

The Red Poll and Farm Conditions

than half that produced by the Guernsey and more than five cents per pound cheaper than the Jersey. We also find the Red Polls away ahead in the average profit per cow. These are not figures of an isolated instance. Red Poll History is replete with just such achievements. I think the Red Poll the most economical producer known, either at the pail or on the block. Also we find the Red Poll Distanced only by one Hereford which beat by the narrow margin of 8-100 of a per cent.

We find that the Red Polls led the following breeds:

The Shorthorns by 8% in average profit per animal.

The Jerseys by 26% in average profit per animal.

The Holsteins by 20% in average profit per animal.

The Guernseys by 49% in average profit per animal.

The Mixed Breeds by 32% in average profit per animal.

The Angus by 26% in average profit per animal.

Fell behind one Hereford 18-100 of a per cent.

Perhaps some of my readers may

The Red Poll and Farm Conditions

think the writer not impartial in his choice and selection of this article of Mr. Kennedys; that he is seeking to make capital of a favorable opportunity. If such is your view you do me wrong. I cited above my reasons for taking this article. I took it because it was extraneous and outside the fold. The Student of Red Polls knows only too well how replete with such victories is the history of the Red Polls, on either side the Atlantic. Here at home in Ohio, the skeptic can find more than food for thought even should he go no farther than the State University and stop with May-Sower No. 2964A12's record. No, I truly believe Red Poll Breeders as a class are more impartial than are other Breeders. Why? Because most of us have served our apprenticeship at the benches of other Breeds. We are young, so to speak. Many, many of us left our pets, that had really become part and parcel of us, to step into Red Poll lines, not when we first had sight and notice of the hand writing on the wall, but only when the interpretation of the writing became a warning so audible we were forced to heed; forced to catch step.

The Red Poll and Farm Conditions

While many of us, as Red Poll Breeders, may be able to "hark back" only too few years, think not, gentle reader, it is so with our choice of Breed, the Red Poll. If you are in doubt or ignorant as to just when Red Poll ancestry begins, read Bede, who died about 700 A. D. or turn to your Encyclopedia Britanica there to find the Suffolk Duns pronounced as **indigenous** to the country. But, returning to our figures cited above, covering cost of production per pound of Butterfat and average profit per Animal; there is another lesson in these same figures. Has it appeared to you? I mean the lesson of proportion. Study it.

There is equally as many Farmers in other States, than in this section of Iowa, that are just as far off the track; that have not yet deciphered, if they have noticed, the hand writing on the wall. How long will it be before they do? Are you, my dear sir, one of the number? Are you seeking enlightenment on the Red Poll? Do you wish it? Is so, apply to the undersigned who will send you an X-Ray on the subject entitled, "Facts and Figures."

FRANK NELSON,
Sec.-Treas.

www.ingramcontent.com/pod-product-compliance
Lightning Source LLC
Chambersburg PA
CBHW062237220526
45471CB00009B/3528